LONGCASE CLOCKS

Joanna Greenlaw

A brass-faced eight-day clock, with phases of the moon, by Evan James of Dolgellau, Gwynedd, c.1755.

Published in 2011 by Shire Publications Ltd,
Midland House, West Way, Botley, Oxford OX2 0PH.
(www.shirebooks.co.uk)

Copyright © 1999 by Joanna Greenlaw.
Shire Library 370.
First published 1999; reprinted 2001.
Transferred to digital print on demand 2011.
ISBN 978 0 74780 417 8

Joanna Greenlaw is hereby identified as the author of
this work in accordance with Section 77 of the
Copyright, Designs and Patents Act 1988.

British Library Cataloguing in Publication Data:
Greenlaw, Joanna
Longcase clocks
1. Longcase clocks
2. Longcase clocks – History
I. Title
681.1'13
ISBN-10 0 7478 0417 6
ISBN-13 978 0 74780 417 8

Cover: *Detail and full case of a mahogany antique longcase clock with moonphase. The clock was manufactured by Lawley & Co., c. 1825. Photograph courtesy of P. A. Oxley Antique Clocks.*

ACKNOWLEDGEMENTS
Photographs are acknowledged as follows: Clifford and Yvonne Bird, page 34 (top); Country Clocks, Tring, pages 13 (right), 26 (left), 28 (left), 37 (all); T. P. Cuss, page 19 (bottom); Joanna Greenlaw, page 7 (top right); G. K. Hadfield, pages 10 (right), 11 (all), 12 (bottom right), 16 (top left), 17 (right), 32 (left); Prescot Museum of Clock and Watch-making, pages 28 (right), 33 (both); Derek Roberts, pages 5, 16 (right), 20 (right), 21 (both), 23 (top left and right), 30 (right), 31,32 (right); Sotheby's, pages 1, 4 (left), 17 (left), 20 (left), 24 (right), 30 (left); Frances Tennant, pages 4 (right), 23 (bottom left), 24 (top left and bottom left), 25 (all), 29 (left), 40 (both); Anthony Woodburn, pages 13 (left), 16 (bottom left), 29 (right); the Worshipful Company of Clockmakers, page 39 (top). The diagrams were drawn by Joanna Greenlaw.

Printed and bound in Great Britain.

CONTENTS

PREFACE

Everyone likes a longcase clock. As the saying goes, every home should have one; and not some reproduction affair, but an old Georgian or Victorian gentleman that has ticked and struck the hours for perhaps a century and a half or more. Such a clock enhances an old house; without one an ancient hall seems incomplete, while in a new one it will offset the brashness of modernity.

And what interesting people the old clockmakers were! In their own day they were the most skilled of local craftsmen, often with a knowledge of mathematics and astronomy rare in country districts. Some were immigrants or refugees, for example Huguenots or Jewish craftsmen fleeing from persecution in Russia, Poland or Germany. Old newspapers are full of the minutiae of their lives: accidents, scandals, robberies, bankruptcies, tragedies and infringements of the law. But the best commemoration of the quietly creative, meticulous labours that formed the fabric of their lives is the clocks themselves.

Left: *A clock by William Holloway of London, c.1710.*

An unrestored dial, showing good examples of detailed world maps, by W. C. J. Nicholas of Birmingham.

THE MEASUREMENT OF TIME

Mankind has always had an awareness of time. When early man began to farm livestock instead of hunt he encountered the seasonal fertility of the animals and the land, and thus the measurement of time began.

The most fundamental clock of all is the earth itself, rotating on its own axis and revolving around the sun in common with the other planets against a background of stars. From the earth it is the stars that appear to move, and the interval between two successive appearances of the same star is 24 hours, a measurement known as sidereal time. The same result can be obtained by observing the sun, when it is called solar time. Because of the relative nearness of the sun and because the earth moves around it as well as rotating on its own axis, there is a slight difference between sidereal and solar time, which can be calculated. Since the earth is tilted by $23^1/2$ degrees, the length of each day (except at the equator) varies throughout the year, the variation being greater towards the poles. In the northern hemisphere the longest day is 21st June and the shortest is 21st December.

The other great indicator of time is the moon, which has regulated human life from the earliest times. The average interval between one full moon and the next is $29^1/2$ days, and before the era of electricity this cycle tended to govern social and business visiting, especially in the country, nights with a full (or almost full) moon being the most favourable for travel. Fishermen and other seafarers found their lives governed by the tides (which depend on the position of the moon); so too did the traders whose goods were transported by sea, since they would want to know when laden ships could get up the river estuaries.

Apart from observing the heavens, the earliest forms of time measurement were of the simplest kind and, in those unpressured days, quite efficient enough. The earliest was probably the graduated candle, followed by various devices based on water dripping from a container through a small hole. The falling water level indicated the passage of time; for instance, a graduated container might empty in twelve hours, when it could be filled up again. Ingenuity soon suggested attaching pointers to floats, and thus the first clock hands came into being. Such devices were known as *clepsydrae*

An eight-day clock by George Graham of London, c.1745. It has a walnut case with caddy top. George Graham was one of the most eminent English members of the Worshipful Company of Clockmakers, who improved (or perhaps invented) the cylinder escapement. He was assistant to Thomas Tompion and, like him, is buried in Westminster Abbey.

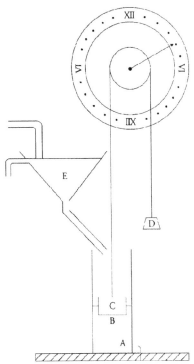

The principle of a clepsydra. The small hole (A) allows the water level in chamber B to fall. Float C drops, thus turning the hour hand (weight D). Reservoir E is refilled when it empties.

and were still in use in the middle ages. The invention of blown glass made the hour-glass possible, and its humble relative the kitchen egg-timer is still popular today.

As an ingenious means of measuring solar time, the sundial was an extremely important early timekeeper. Whether fixed on the face of a public building or horizontally on a pillar in a square, churchyard or manor garden, the sundial was ubiquitous from the middle ages onwards. Until the seventeenth or eighteenth century, the average community, except perhaps the squire, vicar and doctor, would have depended on a public sundial on either a church or the town hall for their time, and on cloudy days they would have had to rely on guesswork. Since the time given by a sundial is solar time it varies according to season (in approximate terms on 12th February the sundial is fourteen minutes slow and on 6th November sixteen minutes fast), but though our forebears suffered numerous inconveniences compared to today they also had many advantages, one of which was not feeling compelled to rush about.

The raised angular piece on a sundial which casts the shadow is called the *gnomon*, and its sloping edge is the *style*. For a sundial to give the correct time the angle of the style has to correspond to the number of degrees of latitude at which the dial is situated. The lines marking the hours are graduated in spacing, being closest together for the hours around noon. Their precise spacing varies according to latitude.

An interesting variation of the sundial, very popular in the eighteenth century, was the 'noon-mark', illustrated in *La Gnomonique Pratique* by Bedos de Celles (1780). When the sun shining through the hole in the disc strikes the vertical line marked on the wall it is noon locally.

The noon-mark, a vertical sundial showing only the 12 o'clock hour line. The spot of light projected by the gnomon crosses the line at exactly noon, and watches could be set by it. This illustration is from the eighteenth-century work 'La Gnomonique Pratique' by Bedos de Celles.

6

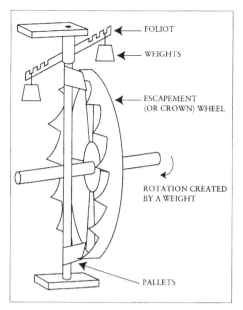

FOLIOT

WEIGHTS

ESCAPEMENT (OR CROWN) WHEEL

ROTATION CREATED BY A WEIGHT

PALLETS

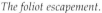

The foliot escapement.

This turret clock at Salisbury Cathedral dates from c.1386 and is said to be the oldest clock in England. It used to be in the central tower and was in use until 1884, but it now stands in the north aisle.

In Europe the first mechanical clocks were made possible by the invention of the *escapement*, which is basically a simple arrangement for letting a toothed wheel rotate one tooth at a time. The earliest form was the foliot escapement, whose action is illustrated in the diagram. The escapement wheel (called a crown wheel) rotates one tooth at a time, and the speed is regulated by the *foliot* – a notched bar which carries an adjustable weight at each end. The nearer the weights are to the centre, the shorter the vibration and the faster the clock will go.

The foliot escapement was a feature of the early turret clocks (the

Below: *An early print depicting a clockmaker's workshop c.1550-60.*

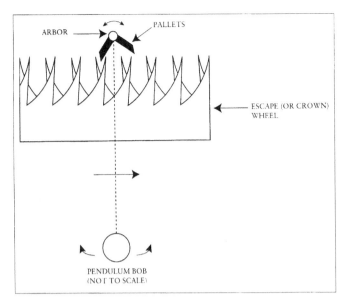

The verge (or crown wheel) escapement.

ARBOR

PALLETS

ESCAPE (OR CROWN) WHEEL

PENDULUM BOB (NOT TO SCALE)

name given to clocks in towers), which became common on church and town-hall towers. A turret clock is defined as a clock where the driving mechanism is at a distance from, though connected to, a large vertical dial. Siting such clocks in a tower combined public visibility with the full drop needed for the driving weight to fall. Turret clocks have simple mechanisms, and once the basic design was established it could be copied by skilled blacksmiths. From these worthy men, who soon began to deserve the name of clocksmiths, sprang the whole race of clock- and watchmakers, culminating in such skilled makers as Joseph Knibb and the incomparable Thomas Tompion (1638-1713), who made clocks for Charles II and now lies buried in Westminster Abbey.

The next development was the verge escapement. The crown wheel is now horizontal, and the *arbor* (axle) carrying the *pallets* has a pendulum attached to it. As before, the escapement rotates one tooth at a time, but the speed is dictated by the periodicity of the pendulum, a function of its length.

THE
ARTIFICIAL
Clock=maker.
A Treatife of
Watch, and Clock-work:
Wherein the Art of
Calculating Numbers
For moft forts of
MOVEMENTS
Is explained to the capacity of the Unlearned.

ALSO THE
Hiftory of Clock-work,
Both Ancient and Modern.
With other ufeful matters never before Publifhed.

By *W. D. M. A.*

LONDON,
Printed for *James Knapton,* at the *Crown* in St. *Paul's* Church-yard, 1696.

HOROLOGICAL
DIALOGUES
In Three Parts
SHEWING
The Nature, Ufe, and
right Managing of
CLOCKS
AND
WATCHES:
WITH AN
APPENDIX
Containing Mr. *OUGHTRED's*
Method for Calculating of Numbers.
The whole being a work very neceffary for all that make ufe of thefe kind of Movements.

By *J. S.* Clock-maker.

London, Printed for *Jonathan Edwin* at the *Three Rofes* in Ludgate-ftreet, 1675.

Left: *'Horological Dialogues' by John Smith, clockmaker and painter (1675). Oughtred's calculations were written in Latin.*
Far left: *'The Artificial Clock-maker' was published in 1696.*

8

Christiaan Huygens (1629-95).

CHRISTIANUS HUGENIUS
natus 14 Aprilis 1629.
denatus 8 Junii 1695.

THE DEVELOPMENT OF THE PENDULUM CLOCK

Christiaan Huygens (1629-95), a Dutch mathematician, is generally credited with having first applied the principle of the pendulum to clocks in 1656, as described in his *Horologium Oscillatorium* (1658/1673). Huygens assigned the right of using a pendulum to a Dutchman, Salomon Coster, in 1657 for a period of twenty-one years. The first use of it by an English clockmaker was in 1658, by Ahasuerus Fromanteel, born in Norfolk of Dutch descent in 1606. Fromanteel's early clocks had short bob pendulums, but in 1659 he produced the first known weight-driven longcase pendulum clock. The ebony case was tall, narrow and architectural – ideal for setting off a burnished brass and gilt dial with cherub spandrels in each corner.

The anchor escapement.

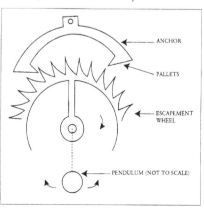

A great advance came around 1675 with the invention of the anchor escapement, which responded to a much smaller arc of motion, thus permitting a longer pendulum with consequently superior time-keeping. This was a key development in the evolution of longcase clocks, which now began to be produced in large numbers. At first they were single-handed clocks (hour hand only) and ran for thirty hours on one winding by a

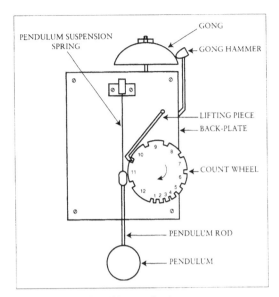

The count-wheel striking mechanism.

rope-and-weight or chain-and-weight arrangement. The mechanism was contained within what is referred to as a bird-cage structure – consisting of a base and top separated by pillars at each corner. The arbors, gear wheels and other moving parts were accommodated by vertical metal strips (pierced to provide pivot holes for the ends of the arbors).

At first the moving parts were made of iron. Later, brass was used increasingly, but not for the arbors, various lifting arms, gong hammers and some other parts. The choice of metal was an economic one: brass was ten times more expensive than iron.

Soon bird-cage arrangements began to be replaced by fitting the mechanism (usually now called the movement) between two vertical brass plates, and this lent itself to the addition of a minute hand. This was followed by a dial or date ring to indicate the day of the month. A second hand came next, together with dials showing the phases of the moon (see next chapter) or the tides.

These improvements coincided with the introduction of arrangements for striking the hour, which were soon supplemented by musical chiming mechanisms. The first of the striking arrangements was the *count wheel* (see diagram). The count wheel is arranged by suitable gearing to rotate once every twelve hours and is provided with slots on its circumference, cut to produce twelve bearing surfaces of varying length corresponding to the numbers one to twelve. The lifting arm rides along these surfaces and releases a striking mechanism, basically a hammer powered by a weight, which strikes a gong. The number of strikes is dictated by the length of the bearing surface, and as

Right and left: *A single-handed brass-faced thirty-hour country clock by William Hoadley of Rotherfield, Sussex, c.1758. It has a plain oak case, a silvered chapter ring and a matted dial engraved with a bird-and-flower motif. The movement is unusual, being of lantern construction and fitted with a rack striking mechanism which can be operated by pulling a cord to repeat the last hour. Mr George Hadfield, by whose courtesy this illustration appears, suggests the intriguing possibility that some crusty old Sussex farmer had a suitable cord running through his bedroom floor so that he could pull the last hour struck without getting out of bed.*

soon as the tail of the lifting arm falls into a slot the striking action is stopped. It is a very simple system but suffers from the disadvantage that the number of strikes can become out of phase with the hour shown on the dial. However, all that has to be done to correct matters is to activate the lifting piece until the count wheel works its way round to the correct position.

The *rack striking mechanism* was invented in 1676 by a gifted amateur, the Reverend Edward Barlow, and possesses the advantage that the number of hours struck is always in phase with the position of the hour hand on the dial. It was also possible for the clock's owner, by pulling a cord, to repeat the last hour that had been struck. This feature was useful at night-time, as the cord could easily be located in the dark.

Left and far left: *An exemplary brass-faced eight-day clock by Benjamin Shuckforth of Diss, Norfolk, c.1725; height 6 feet 10 inches (2.083 metres). The case is of honey-coloured oak.*

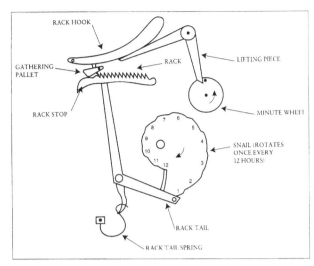

Left: *The rack striking mechanism.*

Below: *An eight-day painted-face clock by John Chambley of Wolverhampton, c.1780; height 6 feet 7 inches (2.007 metres). The case is of oak and mahogany.*

The principle of the rack striking system is shown in the diagram. The snail wheel rotates once every twelve hours, and the twelve different distances between its outer rim and the centre dictate how far the rack tail can be driven upwards by the rack spring and consequently how many teeth on the rack the gathering pallet can pick up. The latter is connected by gearing to a wheel carrying pins which trip a hammer, causing it to strike the gong. Each revolution of the gathering pallet produces one strike on the gong, and the action is terminated when the end of the gathering pallet comes to rest on the rack stop. The striking action is set off by the minute wheel, which rotates once per hour. A pin on this wheel pushes the tail of the lifting piece, which releases the striking mechanism, and at the same time the rack tail comes to rest on the appropriate section of the snail wheel rim, driven there by the rack spring. It first goes into a 'warning position', dictated by a pin on a wheel known as the 'warning wheel', and then, at the hour, the lifting piece drops and the striking commences. Because the hour hand is fixed to the snail it is always in phase with the number of blows struck by the gong, governed by the distance of that part of the snail rim from the centre.

By 1700 most important towns had at least one clockmaker. In country districts thirty-hour clocks were superseding eight-day clocks, as they were cheaper to make. The eight-day clock is characterised by two weights which drive the winding barrels for the going and striking trains (mechanisms) respectively. A thirty-hour clock

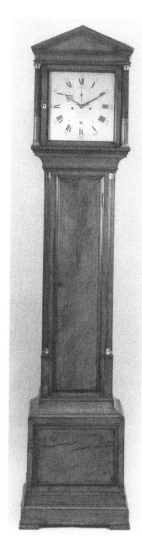

has sprocket wheels instead of winding barrels, both trains being driven by an endless cord powered by one weight. This device is named after Christiaan Huygens, who invented it. In earlier thirty-hour clocks rope was used; later clocks had a chain fitting over the sprockets. Thirty-hour clocks require winding each day, while eight-day clocks, as their name implies, are wound once a week.

From about 1675 month-going clocks appeared. The taller the clock case, the longer the clock will keep going, because of the weight drop, but the only way to make a clock of a practical size keep going for as long as a month is to add an extra wheel in the train (which means that the clock has to be wound up anticlockwise). Usually such clocks were of finer quality, and although they were normally provided with heavier weights, many would run on a standard 14 pound one. A year-going clock required two or three extra wheels, so it might be wound in either direction; but invariably a heavier weight was needed.

The famous clockmakers Thomas Tompion and Daniel Quare (1649-1724) were especially well known for their fine year-going clocks.

Chiming and musical clocks took various forms, the simplest being the 'ting-tang', which chimed 'ting-tang' on two bells, once at quarter past, twice at half past, and so on. Musical clocks with strikers lifted by a rotating pin-barrel could play a selection of chimes on the hour, depending on the design. These were selected

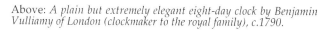

Above: *A plain but extremely elegant eight-day clock by Benjamin Vulliamy of London (clockmaker to the royal family), c.1790.*

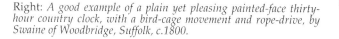

Right: *A good example of a plain yet pleasing painted-face thirty-hour country clock, with a bird-cage movement and rope-drive, by Swaine of Woodbridge, Suffolk, c.1800.*

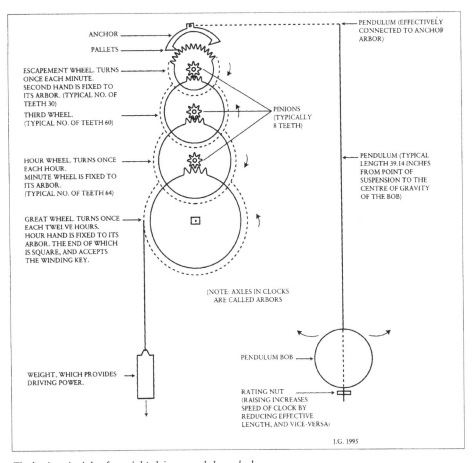

The basic principle of a weight-driven pendulum clock.

by a lever, which could switch to 'silent'. Such 'strike/silent' clocks required as many as twenty-four bells, and some played a different melody each day, with a hymn on Sunday.

The Worshipful Company of Clockmakers of the City of London was formed in 1631 and laid down regulations regarding the number of apprentices a clockmaker might take and the inspection of work (which, if of inferior craftsmanship, was destroyed publicly). It exercised strict control in the cities and towns. In country areas craftsmen were left largely to their own devices, the main restriction being that an apprentice, having completed his time, could not operate within a given distance from the master's abode; he was also banned from areas governed by the Company.

In 1797 an Act of Parliament imposed a duty of two shillings and sixpence on silver watches, ten shillings on gold watches and five shillings on clocks. The Act was repealed in 1798, but not before it had

Left: *Thomas Tompion (1638-1713). He was the inventor of the cylinder escapement for watches (1695) and perfected the rack striking mechanism.*

Below: *A print of 1748 showing a clockmaker at work (from 'The Universal Magazine').*

Above: *A high-quality brass-faced clock by John Shaw of Holborn, London, c.1685. John Shaw was master of the Clockmakers' Company from 1712 to 1715.*

Above: *An interesting eight-day brass-faced clock with a 'strike/silent' mechanism by William Scafe of London, c.1740. The Honourable B. Fairfax, writing to his nephew in 1727, noted: 'One William Scafe, a watchmaker born at Bushey near Denton [Yorkshire], served his time to his father, a blacksmith, but [is] now perhaps the most celebrated workman in London and Europe.' William Scafe rose to become master of the Clockmakers' Company in 1749.*

Left: *A Queen Anne musical clock by Christopher Gould of London, c.1700, in a magnificent walnut case.*

Above: *A fine musical clock by John Webb of Ubley, Somerset, c.1720. The 13 inch (330 mm) arched dial has apertures for month, day and deity. The three-train musical mechanism plays on a nest of eight bells, with striking on a larger bell. The mahogany case has broken-arch cresting and chequered stringing throughout.*

Right: *An eight-day brass-faced clock by Richard Kenfield of Winchester, c.1795; height 6 feet 5 inches (1.956 metres). It has a mahogany case, a 'strike/silent' mechanism and a silvered chapter ring.*

17

A wheel-cutting engine, c.1750.

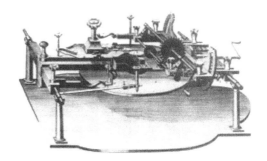

driven many clockmakers out of business.

For clockmakers, as for many others, the nineteenth century was a period of rapid change. By the start of the century many were buying in mass-produced parts, mostly manufactured in Birmingham, and by the 1830s it was usual for the entire mechanism to be bought in. The clockmakers became assemblers and retailers, though they still possessed the basic skill to make any parts they needed. Economics drove their changing practice, and finally their world was altered for ever by the arrival *en masse* of imported clocks from America, Germany and France.

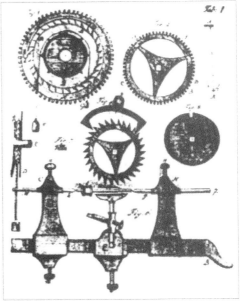

Above: *Horological tools and wheels, c.1724.*

Left: *A clockmaker's trade card, c.1750.*

18

TAX-OFFICE.[1]

Duties on CLOCKS and WATCHES.

To

IN purſuance of an Act of Parliament paſſed in the 37th Year of His preſent Majeſty's Reign for granting certain Duties on Clocks and Watches, you are hereby required to prepare and produce, within Fourteen Days from the Date hereof, a Liſt of all the Clocks, or Timekeepers uſed for the Purpoſe of Clocks, placed in or upon your Dwelling-houſe, and alſo of all Watches, and Timekeepers uſed for the Purpoſe of Watches, kept and worn, or uſed *(after the 1ſt Day of Auguſt* 1797*)* by you, or by any Perſon or Perſons dwelling in your Houſe; which Liſt muſt be ſigned by yourſelf, and muſt contain the Chriſtian and Surname of every ſuch Perſon or Perſons, whether they are liable to the ſaid Duties or not, and whether ſuch Perſon ſhall be of your Family or an Inmate or Lodger therein, ſpecifying the Number of Clocks and Watches, according to the Schedule on the other Side.

The Penalty for not making a Return, and alſo for every Clock or Watch omitted, is *Ten Pounds.*

Dated at *Lamont* this *28* Day of *September* 1797.

Joſeph Morton Aſſeſſor.

Note, Perſons occupying Houſes not aſſeſſed to the Duties upon Houſes and Windows of the 6th Geo. III. and Inhabited Houſes of the 19th Geo III. having only one Clock or one Silver or Metal Watch, are exempted from the Duties thereon.

And alſo Perſons occupying Houſes, together with Offices, not rated at more than Ten Windows, are exempted from the Duty on one Clock, provided the Movements are made of Wood, and the Price or Value of which is not more than Twenty Shillings.

No Servant in Huſbandry dwelling with his or her Maſter or Miſtreſs in any Houſe exempt from the Duties on Houſes of the 19th Geo. III. nor any Non-commiſſioned Officer or Private in His Majeſty's Army, or in the Marines, or the Militia, nor any Seaman in His Majeſty's Navy, or employed in the Merchants Service, are liable to the above-mentioned Duties.

Receipt for tax paid on a clock, 6th June 1798.

Right: A fine eight-day brass-faced clock with a moon dial and quarter chiming on either four or eight bells, c.1775.

Below: An eight-day painted-face clock with a moon dial and map spheres, by Moore of Birmingham, c.1800.

CLOCK DIALS AND HANDS

Until about 1780 clock dials were made of brass, with suitable engraving filled with black wax. The brass was an alloy of about seventy per cent copper and thirty per cent zinc, which resulted in a light colour, and was produced by melting the metals and pouring them into mould trays formed from sand. Brass sheets made in this way were often uneven in thickness, with many flaws. The cooled sheets were hammered, to flatten and harden them, then filed and scraped, the final polish being achieved by rubbing down with powdered pumice and water. Silvering involved rubbing the surface with a paste of silver nitrate, salt and water.

Dials can throw light on the period and authenticity of a clock, the size of brass dials being a good indicator of date. Early dials were small, 9 or 10 inches (229-254 mm) in diameter, the size increasing to 11 or 12 inches (279-305 mm) by the 1770s and 13 inches (330 mm) in the nineteenth century. Decorative brass casings called *spandrels* were used to fill in the gaps produced by fitting a dial to a square or arched format; their patterns also provide a useful aid to dating. Forged

Above: *The dial of a George III musical clock by Samuel Smith of London. The case is mahogany, and it has 'strike/silent' and 'music/silent' mechanisms. The painted scene in the arch incorporates automata: figures playing the harpsichord, violin and lute.*

Right: *A fine small brass-faced eight-day clock, showing the phases of the moon, by Benjamin Bold of London, c.1770; height 7 feet 3 inches (2.210 metres). It has a mahogany case with glazed sides to the hood.*

signatures on dials are far from rare; however, if one knows what to look for, a spuriously modified dial is not hard to identify.

Painted dials were usually made of wrought iron, though occasionally they were of brass, in which case they were generally round. Early moon dials (see below) were brass until about 1825, after which iron was used. Before the painting process, brass dial feet were riveted in place and smoothed over on the dial front. Red oxide paint was used as a primer, and over this as many as eight coats of paint were laid, each being rubbed down, until the top coat was applied with a very fine brush. Painted dials are sometimes referred to as enamelled dials; this is incorrect, as they were almost invariably painted.

A sub-frame called a 'false plate' was usually used to fit the dial to the movement. Early false plates were made of cast iron, with the makers' names cast into them. Since cast iron is brittle, the false plate could easily fracture if dropped on the floor; for this reason, and also because of cost, sheet iron plates were in general use by the 1830s.

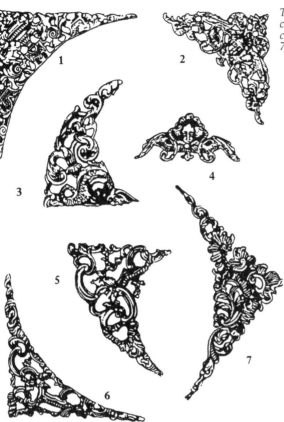

Typical spandrels: 1, urn, c.1750; 2, cherub, c.1700; 3, dolphin, c.1750; 4, cherub, c.1670; 5, c.1775; 6, c.1775; 7, leaf and scroll, c.1775.

The best-known dial maker is James Wilson of Birmingham, whose period of production ranged from about 1777 to 1809. His output was prolific, running into many thousands. Frances Tennant (see 'Further reading') remarks that no two of his dials are exactly alike. Production at his workshop was divided into stages, and teams of artists were employed. An arch over the dial was the ideal place for the main decorative motif, and this would be reflected in the four spandrels. The most popular theme was the four seasons, usually represented by a lady in each corner, appropriately dressed for the season or holding suitable emblems, such as spring flowers or sheaves of wheat.

The introduction of the anchor escapement permitted the inclusion of a second hand and subsequently of moon dials. The latter were usually set in the arch above the clock face, though sometimes, in the case of a square dial, in an aperture. A small moon dial in a round hole was known as a penny moon. A moon dial contained within the arch of an arch-dial clock could be larger and carried a representation of the moon, which, as it rotated, indicated the phases of the moon. There were usually two moons painted on each moon dial, with seascapes or

Above and right: *A fine eight-day clock by John Fairey of Ratcliffe Highway, London, c.1800. It has a mahogany case and a one-piece dial, engraved and silvered, with a 'rocking-ship' automaton.*

Below: *A dial by Joseph Symcock of Nantwich, Cheshire, c.1815, with a break-arch hunting scene.*

Above: *A typical moon dial.*

Right: *A William III brass-faced clock by Christopher Gould of London, c.1700. It has an 11 inch (279 mm) dial with an engraved Tudor rose at the centre and cherub and leaf spandrels. The case is of walnut. Christopher Gould was a well-known maker of high-quality clocks. Although declared bankrupt in 1706, he served as a beadle from 1713 until his death five years later.*

Below: *A dial by Xaver Ganz of Swansea, c.1866. It is a typical break-arch painted dial of the period, depicting a farmyard scene.*

Far left: *A dial by R. B. Pratt of Haverfordwest, Pembrokeshire, c.1850, decorated with a high-quality coastal scene.*

Left: *A break-arch dial by John Gibson of Saltcoats, Ayrshire, c.1830. Its luminous painted decoration illustrates the theme 'a farmer visits a friend'.*

Far left: *A large break-arch dial, with lion's eyes automata, by David Elias of Llanerchymedd, Anglesey, c.1850.*

Left: *A dial by John Barnsdale of Burnham, probably early nineteenth-century, with a break-arch country scene.*

landscapes, etc, between the moons. In the days before street lighting a moon dial was an important feature, because whenever possible people regulated their night-time travelling by it. Because the moon governs the tides, the moon-dial principle lent itself to indicating tide levels, so tidal dials were specially useful to shipowners and merchants in sea-ports. If there was an arch but no moon dial, a 'strike/silent' dial might be placed in the arch. Another possibility was simply some decorative motif or, later, painted scenes and even automata, such as rocking ships or swans.

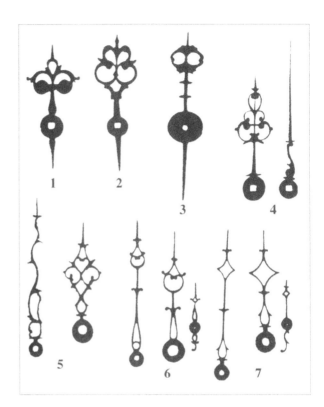

Above: *Typical clock hands: 1, 1650-75; 2, 1700-50; 3, 1675-85; 4, 1750-75; 5, 1775-1825; 6, 1800-75; 7, 1790-1825.*

Left: *An eight-day painted-face clock by Bromley of Horsham, Sussex, c.1815. Its arched dial has a floral motif.*

Among other forms of ornament, fruit and flower decoration was very popular. Favourite decorative motifs were double roses, strawberries and anemones.

Roman numerals were used until about 1800, when Arabic numerals came in, but by 1825 Roman numerals were again fashionable. On painted dials Arabic numbers were invariably used for second-hand and date rings.

Clock hands form a study in themselves, and because their patterns clearly changed over the years they provide a useful guide to period. Early clocks had one hand, equipped with a tail that as well as having an aesthetic function, provided leverage for setting it. Hands were made of steel until the early 1800s, when brass generally superseded it, permitting more elaborate piercing and chasing.

CLOCK CASES

A longcase clock was often the most expensive and decorative piece in a house. Its appearance was of prime importance, whether it was to stand in a cottage or a mansion.

In the sixteenth and seventeenth centuries oak was the commonest material used, though the London makers tended to prefer ebony – or wood stained to resemble it. Ebony was very common in early longcase

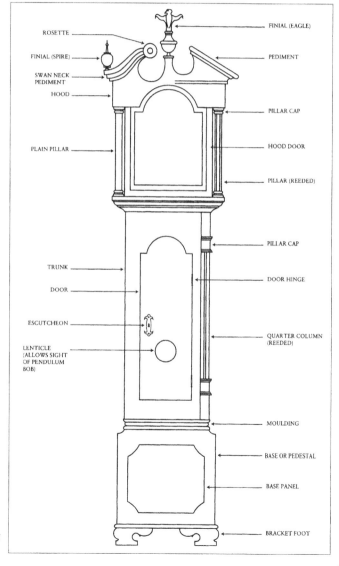

The various parts of a clock case.

Left: *An eight-day brass-faced clock by Abraham Rusted of Litlington, Cambridgeshire, c.1720. It has a square dial with silvered chapter ring. Its high-quality oak case has been cut down at the bottom, probably to accommodate a low ceiling.*

Right: *An eight-day clock by William Simcock of Prescot, c.1770. The case is of burr oak.*

clocks because it showed off a brass and silvered dial to advantage, but it ceased to be so after about 1700. Walnut became popular in the 1650s and continued in favour until the 1750s, when it became expensive owing to scarcity. Mahogany was available in the late seventeenth century but little used, being subject to a heavy import duty, but when the duty was removed in 1721 it rapidly became popular. It was close-grained and easy to work, lending itself to mouldings, fretting and carving. In country districts oak was almost universal, only being superseded by mahogany from the 1800s onwards. By then wide deal boards were being imported in quantity from the Baltic, and as the

A decorated pendulum bob.

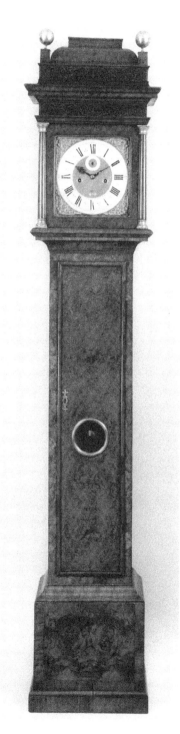

Right: *An elegantly proportioned Queen Anne eight-day clock by Charles Gretton of London, c.1710. The case is veneered in burr walnut and has a lenticle for viewing the pendulum.*

nineteenth century wore on the use of basic softwood structures with mahogany or sometimes satinwood veneers became widespread. Pine cases were the cheapest of all and were usually painted, to protect them from rot and woodworm.

A *lenticle glass* was a circular window in the trunk door, through which the swinging pendulum could be viewed. It was popular only for a short period – 1675-1750, which makes it a good dating guide. At about the same time glass windows were often fit-

Far left: *A George II clock by George Booth of Manchester, c.1750, with a revolving 'Halifax' moon sphere in the dial arch. The case is of walnut, with domed caddy cresting and gilt wood finials. The trunk door is cross-banded, with an inlaid wheat-ear border.*

Left: *A fine small eight-day brass-faced clock by Edmund Brewer of London, c.1765. It has a mahogany case, a matted dial with fine engraving and a 'strike/silent' mechanism.*

ted to the sides of the hood so that the movement could be seen.

In very broad terms, case-making can be summarised as falling into three overlapping periods: before 1700, 1700 to 1800 and 1800 onwards. The earliest period is characterised by dark colours, oak giving way to walnut in the 1650s. In these early clocks the cases were narrow, with very narrow doors. The hood pillars were integral with the door frame. After the arrival of Huguenot refugees in the late seventeenth century extensive marquetry and fretwork were introduced. Cross-banding (marquetry done in a zigzag fashion) became popular, together with stringing (straight, narrow lines of decorative inlay work). The materials used were hardwoods, such as holly or fruit wood, and bog-oak, favoured for its blackness. Back boards were of oak, often roughly sawn or finished with an adze. Hoods were removed by lifting up (by the mid 1700s this had given way to the slide-forward system universal today). They were flat-topped at first and later had 'caddy' tops (so called for their resemblance to the tea caddies popular at the time).

In the eighteenth century hood pediments became increasingly architectural and were often decorated with water-gilded wooden case ornaments or brass-work pillar caps and finials. Hood pillar columns became separate, and quarter columns – either plain or fluted – were added to the case trunk. Mouldings became more elaborate and case trunks wider. During the middle of the eighteenth century London

An eight-day painted-face clock by David Somerville of St Ninians near Stirling, Scotland, c.1805-20. The mahogany case contains a note written by Hugh Scott-Riddell (1796-1870), a minor Scottish poet, presumably then its owner.

Right: *A plain but attractive eight-day painted-face arched-dial clock by Harris of Canterbury, c.1800, height 7 feet 3 inches (2.210 metres). It has a mahogany case, with the sides of the hood glazed, and a fighting ship scene painted in the dial arch.*

Left: *An eight-day painted-face moon-dial clock by Joseph Durward of Edinburgh, c.1800. The case is of mahogany.*

Below: An eight-day musical clock by Joseph Finney of Liverpool, c.1760. It plays seven tunes and indicates the phases of the moon and the position of the sun in the sky. The door contains a mercury barometer. Joseph Finney, who was also an architect, was made a freeman of Liverpool in 1732 and died in 1772.

Above: An eight-day clock by John Wyke of Liverpool, c.1756, with central seconds hand. The case is of mahogany and is a good example of the 'Lancashire Chippendale' style. Wyke was trained as a toolmaker and tool merchant. He was also reputedly Liverpool's first banker. He moved from Prescot to Liverpool in 1758 and died in 1787.

Right: *Clockmaker's billhead, 1822.*

Below left: *An advertisement for Xaver Ganz, a Swansea watch- and clockmaker.*

Below right: *Xaver Ganz and his family.*

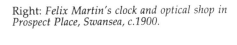

Right: *Felix Martin's clock and optical shop in Prospect Place, Swansea, c.1900.*

makers embellished their cases with Chinese or Japanese lacquer work in brilliant colours, such as vermilion, emerald green, red and blue, with various designs and motifs that were raised and gilded. Originally these cases were sent to the Far East for the work to be carried out, but increasingly imitations were done by English craftsmen, giving rise to the term 'japanning'.

After 1800 basic softwood structures were covered by various veneers – mahogany and satinwood being the most popular. Cases so made were light for their size, though not as robust as their predecessors. The Victorian taste for embellishment led to cases becoming wider, taller and covered in extravagant mouldings, turnings, marquetry and stringing. Case doors particularly lent themselves to decoration by graining and marquetry designs of classical urns, star-shaped inlays or flower arrangements. At the cheaper end of the market, plain pine cases were popular, invariably stained or painted.

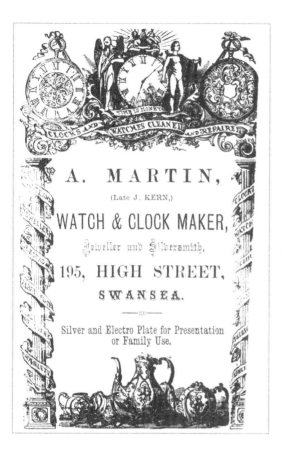

A. MARTIN,

(Late J. KERN,)

WATCH & CLOCK MAKER,

Jeweller and Silversmith,

195, HIGH STREET,

SWANSEA.

——:o:——

Silver and Electro Plate for Presentation
or Family Use.

*Two advertisements for Swansea
clockmakers.*

B. LEVINBERG,

PRACTICAL

English & Geneva Watch Maker, Jeweller, etc.

45, HIGH STREET, SWANSEA.

(Opposite the " Adam and Eve.")

B. L. having had many years experience in some of the largest Establishments in England
and abroad, can confidently assure the Public that all Orders he may be favoured with will
be executed in the best style of Workmanship at moderate prices.

SETTING UP AND MAINTENANCE

It is essential for a longcase clock to be stable, and this is best achieved by screwing it to the wall. Usually old clocks have a hole in the back, but if necessary make one. It will prevent disasters (clocks have been known to fall on their faces or from a landing), and the case can then be polished without the movement being disturbed by undue vibration.

All pendulum clocks require 'setting in beat' (setting so that the ticking sounds even). If a clock is out of beat the intervals between the ticks are alternately long and short, so the ticks seem to come in pairs.

The *crutch*, made of soft iron, can be bent by hand, the aim of which is to alter its position in relation to the escapement. By this means, using a system of trial and error, the clock can be put in beat. It is a process that often requires patience, but providing the mechanism is working correctly the optimum point can always be found. It is essential that no strain be placed on the crutch bearing, so pressure must be applied by the fingers in opposite directions.

When transporting pendulum clocks, always remove the pendulum, otherwise the suspension spring which supports it can break (or become twisted, which gives rise to many problems).

When winding a longcase clock it is good practice to help the weight up gently with one hand, taking care not to allow the line to become slack so that it fails to coil round in the grooves on the winding barrel or jumps off and becomes entangled round the winding barrel spindle. It is best to keep a clock going by winding it in good time. If the clock stops because it needs winding never rotate the hands backwards or whizz them round to set the clock to the correct time. Reset the time by turning the minute hand gently clockwise and allow each striking or chiming sequence to work itself through before moving the hand on to the next strike or chime. However, the best procedure is to wait until about ten minutes before the time shown on the dial and then start the clock by pushing the pendulum and moving minute hand the short distance required.

To make the clock go faster, rotate the regulating nut at the bottom of the pendulum so that the pendulum bob rises, making the pendulum effectively shorter. To make the clock go more slowly, reverse the process.

For the owners of thirty-hour clocks who do not want the striking it is useful to know that if the striking mechanism is disconnected the clock will run for several days on one winding. This is

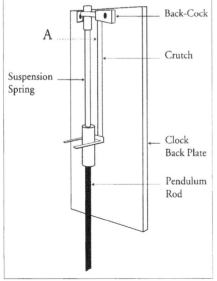

A

Suspension
Spring

Back-Cock

Crutch

Clock
Back Plate

Pendulum
Rod

Crutch adjustment to set a pendulum clock in beat. To do this, pressure must be applied at A with the fingers.

a simple adjustment but should be carried out by someone who understands what is required.

A longcase clock is a mechanical device with bearings that require occasional oiling with fine clock oil. However, excessive oiling is very harmful, attracting dirt which clogs and wears the mechanism. A small drop with a fine artist's brush is all that is necessary.

Never let anyone loose with an oil-can on a clock. In some parts the power is transmitted by friction, which is destroyed by oiling – imagine greasing the clutch plates of a car! If this precaution is not observed, the clock may have to be dismantled to remove the oil, which could prove costly.

Three views of an eight-day clock movement by William Hardeman of Bridge, Kent, c.1810, with a 'dead-beat' escapement. Unusually, the dial shows the date and the day of the week and originally had a central seconds hand.

BUYING AND SELLING CLOCKS

Buying and selling clocks can be a risky business financially. The safest though most expensive way – but probably the best – is to buy from an established dealer who will give at least one year's guarantee plus an assurance of authenticity. When buying a clock from a source not offering a guarantee, beware. Clocks may not be what they seem to the lay purchaser. This applies particularly in auction rooms, where the principle of *caveat emptor* may be applicable. Professional (or knowledgeable amateur) advice on viewing day may save a costly disappointment.

'Marriages' – where a movement has been put into another case – are acceptable, unless either movement or case is of higher quality, but both must be of the same style and period. There is also the danger of excessive restoration, sometimes to a level that destroys the purported value of the clock. Massive restoration can disguise faking, along the lines of the antique furniture 'restorer' who turns three and a half Hepplewhite chairs into a set of six 'restored' chairs. Be especially careful in looking at dials, particularly painted ones. These might have had the name of a good local maker added to them, sometimes very convincingly. Brass dials can be suitably aged reproductions, blatantly obvious to the expert but misleading to the uninitiated.

Above all, the movement may be faulty or so worn that it needs a great deal of money spent to make it reliable. The fact that a clock ticks for a few minutes in a saleroom does not necessarily mean that it is in good order. The striking can be checked by moving the minute hand round slowly clockwise, allowing each hour to strike before going on to the next; in particular make sure that it strikes correctly at eleven, twelve and one.

It may be acceptable to buy a clock described as 'in need of a bit of restoration', if the price is right and a good restorer is on hand who will carry out the work at a reasonable charge, but make sure that this can be done before agreeing to the purchase.

Selling also has its hazards. Dealers need to make a profit and will naturally offer less than the amount for which they hope to sell the clock. For sale by auction, the great London and provincial auction houses are best. In general, though, to avoid potential pitfalls, a professional valuation is a wise precaution when selling a valuable clock.

FURTHER READING

Baillie, G. H. *Watch and Clockmakers of the World* (Volume I). NAG Press, Colchester, first edition 1929 (reprinted 1966).

Darker, Jeff, and Hooper, John. *English 30 Hour Clocks, Origin and Development 1600–1800*. Penita Books, 1997.

de Carle, Donald. *Watch and Clock Encyclopaedia*. NAG Press, Colchester, first edition 1950 (reprinted 1995).

Gazeley, W. J. *Watch and Clock Making and Repairing*. Robert Hale, London, first edition 1953 (revised edition 1993).

Loomes, Brian. *Brass Dial Clocks*. Antique Collectors' Club, Woodbridge, 1998.

Loomes, Brian. *Painted Dial Clocks*. Antique Collectors' Club, Woodbridge, 1994 (reprinted 1996).

Loomes, Brian. *Watch and Clockmakers of the World* (volume II). NAG Press, Colchester, 1976 (reprinted 1992).

Penman, Laurie. *Clock Repairer's Handbook*. David & Charles, Newton Abbot, 1986 (reprinted 1998).

Tennant, Frances. *Longcase Painted Dials*. Robert Hale, London, 1995.

The coat of arms of the Worshipful Company of Clockmakers.

ESTABLISHED 1805.

CLOCK MANUFACTORY,
SOHO STREET, WANDSWORTH
Near BIRMINGHAM.

WILLIAM FREDERICK EVANS
(Successor to JOHN HOUGHTON,)
MANUFACTURER OF
MARINE, CARRIAGE & SKELETON LEVER CLOCK
ALL KINDS OF SPRING AND WEIGHT CLOCKS,
Brass Castings, Forge Work, Pinions, and Clock Materials of ev description.

A nineteenth-century advertisement for a Birmingham clockmaker.

PLACES TO VISIT

British Horological Institute, Upton Hall, Upton, Newark, Nottinghamshire NG23 5TE.
Telephone: 01636 813795. Website: www.bhi.co.uk
The Clockmaker's Museum, Guildhall Library, Aldermanbury, London EC2V 7HH.
Telephone: 020 7332 1868.
Dorset Collection, Mill House Cider Museum, Owermoigne, Dorchester, Dorset DT2
8HZ.Telephone: 01305 852220. Website: www.millhousecider.com/clocks.html
Fitzwilliam Museum, Trumpington Street, Cambridge CB2 1RB.
Telephone: 01223 332900. Website: www.fitzwilliam.cam.ac.uk
Manor House Museum, Honey Hill, Bury St. Edmunds, Suffolk IP33 1HF.
Telephone: 01284 757076. Website: www.stedmundsbury.gov.uk
Museum of the History of Science, Broad Street, Oxford OX1 3AZ.
Telephone: 01865 277280. Website: www.mhs.ox.ac.uk
Prescot Museum of Clock and Watch-Making, 34 Church Street, Prescot, Merseyside
L34 3LA.Telephone: 01551 430 7787. Website: www.prescotmuseum.org.uk
Science Museum, Exhibition Road, South Kensington, London SW7 2DD.
Telephone:020 7942 4454 or 4455. Website: www.nsmi.ac.uk
Tymperley's Clock Museum, Trinity Street, Colchester, Essex CO1 1JN.
Telephone: 01206 282931.
Victoria and Albert Museum, Cromwell Road, South Kensington, London SW7 2RL.
Telephone: 020 7942 2000. Website: www.vam.ac.uk

*A brass break-arch dial of c.1775, by
a Chester clockmaker, with an Adam
and Eve scene in the silvered arch.*

*A moon-dial clock, with upside-
down maps, by Wignall of
Ormskirk, Lancashire, c.1825.*

40

Printed and bound by CPI Group (UK) Ltd, Croydon, CR0 4YY

11/10/2024

01043560-0002